METHYLENE BLUE SIMPLIFIED

How methylene blue is extracted, Produced, and used in medicine

Marvin Stoney

copyright@2024 Marvin Stone, all right reserved, no part of this publication should be reproduce in any form or means without prior written permission from the copyright holder.

Chapter One 4
How methylene blue is extracted, Produced, and used in medicine 4

Chapter Two 8
How Methylene Blue is made chemically and what its features are 8

Chapter Three 10
The steps and methods used for extraction 10

Chapter Four 21
Environment and Safety Consideration: 21

Chapter Five 28
Drug pharmacokinetics and drug pharmacodynamics 28

Chapter Six 43
Diagnostic uses in medicine 43

Chapter Seven 48
Therapeutic applications across different medical fields 48

Chapter Eight 54
Comparative analysis with alternative treatments 54

Chapter Nine 70
Tests and trials in hospitals 70

Chapter One

How methylene blue is extracted, Produced, and used in medicine
What does methylene blue mean?

The molecular formula for methylene blue is $C_{16}H_{18}N_3SCl$. It is a heterocyclic aromatic molecule. Depending on how oxidized it is, it looks like a dark green or blue solid. It is often used as a dye and a medicine. It can be used in medicine as a diagnostic stain, an antiseptic, and a treatment for methemoglobinemia and some diseases, among other things. Researchers are also looking into whether methylene blue can protect neurons and kill malaria parasites.

How important is methylene blue

1. Medical Use

- **Treatment of Methemoglobinemia:** It lowers blood levels of methemoglobin that are too high.

- **Diagnostic Tool:** Methylene Blue is used to color structures in histology and bacteria so that they can be identified under a microscope.

- **Antiseptic:** It can be used to treat wounds and urinary tract diseases because it is an antiseptic.

- **As an antimalarial:** It has shown promise in some tests as an antimalarial.

2. Purposes of Research:

- **Properties that protect neurons:** Studies show that it might have benefits that protect neurons and could be used to treat neurodegenerative diseases.

- **Cell Biology:** It is used to study mitochondrial activity and cellular respiration in cells.

3. Industrial uses:

- **Dyes:** Methylene Blue is used as a dye in the paper and textile businesses.

4. Animal health care:

- It is used in veterinary medicine for the same reasons it is used in human medicine. For example, it is used to treat methemoglobinemia in animals.

Chapter Two

How Methylene Blue is made chemically and what its features are

Chemicals Structure:

- Methylene Blue has a complicated aromatic structure with different types of rings.

- There are thiazine colors in this group.

- The formula for molecules is $C_{16}H_{18}N_3SCl$.

It Features:

- **how does it Look:** Depending on its oxidation state, methylene blue usually shows up as dark green or blue crystals or powder.

- **Solubility:** this can be mixed with water and ethanol.

Stability: It's stable in normal situations, but it can be affected by light.

- **pH Sensitivity:** In alkaline conditions, it changes color to blue, and in acidic conditions, it changes color to nothing.

- **Redox properties:** can accept and give electrons, which is why it is used in many biological tests as a redox indicator.

- **Absorption Spectrum:** It takes in light in the visible range and turns blue when oxidized and green when reduced.

Chapter Three

The steps and methods used for extraction

Getting Methylene Blue out of its solution or mixture usually means using different methods to separate it. Here are some popular ways to get Methylene Blue out of something:

1. Precipitation:

• Methylene Blue can be taken out of its water solution by precipitating it with the right substances, like acids or salts. Changing the pH can also help snowfall happen.

2. Filtration:

• Filtration is often used to separate Methylene Blue from solid impurities or precipitates that form when it is made or cleaned.

3. Getting crystallized:

• Crystallization is the managed growth of Methylene Blue crystals from a liquid. This method is used to clean Methylene Blue and get it in the form of a solid.

4. Distillation:

If Methylene Blue makes an azeotrope or is volatile under certain conditions, distillation can be used to separate it from the reaction mixtures or solvents it is mixed with.

5. Liquids Extraction:

• In this method, Methylene Blue is taken out of an aqueous solution and mixed with an organic liquid. The organic phase is then separated.

6. Chromatography with Ion Exchange:

• Ion exchange resins can specifically bind and release Methylene Blue ions from a solution, which makes it possible to clean it up and make it stronger.

7. Adsorption:

• To get rid of Methylene Blue in water, it can be adsorption onto solid supports like activated carbon or silica gel.

Environmental consideration in production

Because making and using Methylene Blue could have negative effects on the world, it is very important to think about these effects when making it. Here are a few important natural things to think about:

1. Getting raw materials:

• Chemical Feedstocks:
Choosing raw materials and

feedstocks that are better for the environment can cut down on waste and emissions.

• **Water Usage:** Making sure that water is used efficiently and that as little trash as possible is released during production.

2. Using less energy:

• **Energy Consumption**: Using technologies and methods that use less energy to lower greenhouse gas emissions and total energy use.

• **Green Energy:** Powering production processes with green energy sources when it is possible to do so.

3. Waste Management:

• **Hazardous trash:** Handling, treating, and getting rid of hazardous trash properly when it is made during the synthesis or purification process.

• **Recycling:** Looking for ways to recycle or reuse waste goods or intermediates to cut down on waste production.

4. Control of Emissions:

• **Air emission:** Using technologies to control and lower the amount of volatile organic compounds (VOCs) and other toxins that are released into the air.

- **Water Pollution:** Keeping bodies of water clean by treating wastewater properly and controlling its release.

5. Compliance with regulations:

- Following local and foreign rules and laws about the environment when making chemicals and releasing them into the environment.

6. Examining the lifecycle:

- Doing lifecycle assessments (LCAs) to figure out how making Methylene Blue affects the environment from getting the raw materials to throwing them away.

7. The basics of green chemistry:

• Using green chemistry ideas like atom economy, solvent selection, and making as little trash as possible in production processes to lower their impact on the environment.

Making Methylene Blue in factories

In the chemical industry, Methylene Blue is usually made through a chain of reactions that begin with aromatic amines. The process of industrial synthesis can be summed up in these words:

1. What You Need to Start:

Aromatic Amine: Most of the time, an aromatic amine like dimethylaniline or aniline is used as the starting material.

2. Oxidation Process:

• The aromatic amine changes when an oxidizing agent like nitric acid (HNO_0) or sodium nitrite ($NaNO_2$) is present. During this step, a diazonium salt middle is made.

3. Adding Reaction:

• In an alkaline environment, the diazonium salt combines with a good coupling agent, like N,N-dimethylaniline. This connection is what Methylene Blue is made of, usually as the chloride salt.

4. Oxidation and Making:

• More oxidation takes place on the intermediate product to make Methylene Blue. This oxidation step is very important for turning Methylene Blue into its final dark green or blue color.

5. Purification:

• The raw Methylene Blue product is cleaned up using filtration, crystallization, or chromatography to get rid of any impurities and reach the right amount of purity.

6. Formulation and Packaging:

• Methylene Blue is usually made into a solid or liquid form that works for its intended use, like coloring or medicine, after it has been cleaned up.

Chapter Four

Environment and Safety Consideration:

- **Waste Management:** Getting rid of and treating waste and by-products made during synthesis in the right way to have the least amount of effect on the environment.

- **Safety Measures:** Putting in place safety rules for working with dangerous chemicals used in the production process.

Methods for quality control and purification

Quality control and cleaning are very important steps to make sure that Methylene Blue meets the requirements for the uses it was made for. These are popular ways to check the quality of Methylene Blue and make it pure:

Checking for quality:

1. Methods of analysis:

• **UV-Vis Spectroscopy:** This method measures how much Methylene Blue is absorbed at different wavelengths to find out how concentrated it is.

• **HPLC (High-Performance Liquid Chromatography):** This method measures and names Methylene Blue and other elements that are similar.

- **Titration:** Standard titrant solutions are used to make chemical reactions that show how pure Methylene Blue is.

- **Spectrophotometry:** This method measures how much light at different wavelengths is absorbed to figure out how much Methylene Blue is present.

2. Physical Properties:

- **Solubility and Melting Point:** Testing these qualities can help make sure that Methylene Blue is what it says it is and that it is pure.

- **Color and look:** Checking visually to see if the color and look are consistent.

3. Analysis of Impurities:

- **Gas chromatography (GC):** This method is used to find and measure impurities that are mobile.

- **Mass Spectrometry (MS):** gives a thorough look at the structure and molecular weight, which helps find impurities.

Methods of Purification:

1. How to filter:

- **Microfiltration:** This method gets rid of solid impurities or particles from Methylene Blue liquids.

- **Ultra filtration:** This method separates bigger molecules from colloids.

2. Crystallization:

Filtration and crystallization: One way to clean Methylene Blue is to control the crystallization process and separate it from other substances that aren't needed.

3. Chromatography:

Column chromatography separates Methylene Blue from other chemicals based on how neutral they are.

- **Thin-Layer Chromatography (TLC):** This is used for qualitative research and to keep an eye on how the purification process is going.

4. Distillation:

- **Fractional distillation:** This method separates Methylene Blue from volatile impurities based on their different boiling points, making it pure.

5. Exchange of Ions:

- **Ion Exchange Resins:** These are used to clean up Methylene Blue liquids by exchanging certain ions with other ions.

6. Absorption:

- **Activated carbon:** takes in organic elements from solutions of Methylene Blue.

Chapter Five

Drug pharmacokinetics and drug pharmacodynamics

Pharmacokinetics and pharmacodynamics are very important parts of knowing how Methylene Blue works and how it affects the body. Here is a summary:

Pharmacokinetics:

1. Taking in:

• Methylene Blue can be given by mouth, through an IV, or on the skin.

• When Methylene Blue is taken by mouth, it is absorbed from the digestive system and reaches its highest level in the blood within one to three hours.

• When Methylene Blue is injected, it is quickly and completely absorbed.

2. Distribution:

The chemical methylene blue moves around the body and even crosses the blood-brain barrier.

• It attaches to plasma proteins and builds up in tissues, mostly in cells that have a lot of mitochondria.

3. How our bodies use energy

• Methylene Blue is mostly broken down in the liver by NADPH-cytochrome P450 reductase, which makes different molecules.

- In tissues, it breaks down into leukomethylene blue, which is an active reduced form. This form is then changed back into methylene blue through oxidation.

4. Excretion:

- The kidneys are the main way that drugs are flushed out of the body. Both the drug itself and its byproducts are flushed out in the urine.

Pharmacodynamics

1. The way it works:

- Methylene Blue is a redox-active substance that helps cells' electron transfer processes happen.

- It can take electrons from other molecules and give them to other molecules, which can change how cells work and how mitochondria work.

2. Effects on the body:

- Treatment for Methemoglobinemia: Methylene Blue changes the ferric iron (Fe^{3+}) in methemoglobin back to ferrous iron (Fe^{2+}), which makes hemoglobin able to carry oxygen again.

- Antioxidant and neuroprotective effects: Because it lowers oxidative stress and speeds up cellular respiration, it has been looked at as a possible way to treat neurodegenerative illnesses.

- Antimicrobial Properties: Methylene Blue can kill some germs and parasites, but scientific research is still going on to figure out how it does this.

3. Clinical Consideration:

- Dosage and Administration: The right amount of Methylene Blue to give a patient depends on the reason for the prescription and their unique needs, while also taking into account any possible side effects or drug combinations.

• Safety Profile: Methylene Blue is usually well-tolerated when used in the right amounts. However, methemoglobinemia or hemolysis can happen in people who are sensitive to it when it is given quickly or in high concentrations.

How things work in Biological systems

There are several ways that methylene blue affects biological processes, which shows how versatile its pharmacological properties are. Here are the main ways that they work:

1. Cycles of Redox:

• Methylene Blue is a redox-active substance, which means it can give and receive electrons. Because of this, it can take part in redox processes inside cells, which affects the metabolism and function of mitochondria.

• In living things, Methylene Blue can change between its oxidized form (Methylene Blue) and its reduced forms (leukomethylene blue and dihydromethylene blue). This helps cells make energy and protect themselves from free radicals.

2. How mitochondria work:

• Methylene Blue works with parts of the mitochondrial respiratory chain, especially Complex I (NADH dehydrogenase), to improve the flow of electrons and the production of ATP.

• Methylene Blue may lower oxidative stress, boost cellular energy consumption, and protect against mitochondrial dysfunction linked to neurodegenerative diseases by making mitochondria work better.

3. Getting rid of methemoglobin:

• One of the most well-known medical uses of Methylene Blue is

to change methemoglobin back into usable hemoglobin. Iron in hemoglobin changes into the ferric (Fe^{3+}) state, which can't join oxygen. This is called methemoglobinemia.
Methemoglobin is changed into hemoglobin by methylene blue, which increases the amount of oxygen it can carry.

4. Effects on microbes:

Methylene Blue can kill germs and parasites because it is an antimicrobial. It's not completely clear how it works in this situation, but it might involve breaking down cell membranes or stopping certain metabolic processes.

5. Properties that protect neurons:

• Studies show that Methylene Blue may protect neurons, possibly by acting as an antioxidant and improving the function of mitochondria. It might protect neurons from oxidative stress, lower the buildup of proteins that can lead to neurodegenerative conditions, and help neurons stay alive.

6. Effects on inflammation:

It has been shown that Methylene Blue can reduce inflammation by stopping the production of inflammatory cytokines and changing immune reactions. This process is one reason why it might be useful as a medicine for inflammatory conditions.

Interactions with parts of cells

Methylene Blue interacts with different parts of cells, changing how they work and adding to its drug-like affects. Here's how Methylene Blue works with important parts of cells:

1. The mitochondria:

- Methylene Blue works with parts of the mitochondria,

especially the electron transport chain complexes. It moves electrons around, which speeds up the production of ATP.

• Methylene Blue helps keep cellular energy production going and supports cellular respiration by making mitochondria work better.

2. The cytochromes:

• Methylene Blue comes into contact with cytochrome enzymes that help cells transfer electrons. As part of cellular redox processes, it can take electrons from reduced cytochromes and give them to oxidized forms.

3. What are hemoglobin and methemoglobin?

• Methylene Blue fixes methemoglobinemia by changing methemoglobin from a Fe^{3+} state to a functional Fe^{3+} state. This makes hemoglobin molecules in red blood cells able to carry oxygen again.

4. Membranes of cells:

• Methylene Blue can interact with the membranes of cells, which could change how well they work and how well they stay together. It may help its antimicrobial effects against bacteria and parasites because of this interaction.

5. Parts of neurons:

• Methylene Blue has been shown to change the release of neurotransmitters and the way synapses work in neurons. It might keep neurons safe from oxidative stress and help neurons stay alive in conditions that cause neurodegeneration.

6. Strings of DNA and proteins:

• Methylene Blue can attach to proteins and nucleic acids (DNA and RNA), changing their shape and how they work. This interaction could change how genes are expressed, how proteins are made, and how cells talk to each other.

7. Intracellular Redox Balance:

• Through its redox-active properties, Methylene Blue helps maintain intracellular redox balance. It can scavenge reactive oxygen species (ROS) and protect cells from oxidative damage, thereby exerting antioxidant effects.

Chapter Six

Diagnostic uses in medicine

Methylene Blue has several diagnostic uses in medicine, primarily due to its staining properties and ability to visualize specific structures or conditions. Here are some key diagnostic applications of Methylene Blue:

1. Histology and Pathology:

• **Tissue Staining:** Methylene Blue is used as a biological stain to enhance contrast and visualize cellular structures in histological and pathological specimens. It can highlight nuclei and other cellular components, aiding in the diagnosis of various diseases.

2. Microbiology:

- **Microbial Staining:** In microbiology, Methylene Blue is used as a vital stain to differentiate and visualize microorganisms. It stains bacterial cells and certain structures, helping microbiologists identify and classify bacteria under a microscope.

3. Methylene Blue Test:

- **In vivo Testing:** The Methylene Blue test is used in certain medical conditions to assess renal function and detect abnormalities in the urinary system. It involves administering

Methylene Blue orally or intravenously and monitoring its excretion in urine to evaluate kidney function and detect potential urinary tract obstructions.

4. Endoscopic Procedures:

• **Visual Enhancement:** During endoscopic procedures, Methylene Blue can be used to enhance visualization of mucosal lesions, detect abnormal tissue growth, or assist in identifying anatomical landmarks. It aids in improving diagnostic accuracy and guiding therapeutic interventions.

5. Ophthalmology:

• **Eye Staining:** Methylene Blue can be applied as a topical dye in ophthalmic examinations to highlight corneal defects, assess tear film dynamics, or detect foreign bodies. It helps ophthalmologists spot and evaluates ocular conditions more effectively.

6. Neurological Tests:

• **Neurodiagnostic Applications:** Methylene Blue has been explored for its possible role in neuroimaging and diagnosing certain neurological disorders. Its ability to cross the blood-brain barrier and interact with neuronal components makes it a candidate for studying brain function and pathology.

Chapter Seven

Therapeutic applications across different medical fields

Methylene Blue has diverse therapeutic applications across various medical fields, due to its pharmacological properties and mechanisms of action

1. Treatment of Methemoglobinemia:

• Methylene Blue is the antidote of choice for treating acquired and hereditary methemoglobinemia. It converts methemoglobin (Fe^{3+}) back to functional hemoglobin (Fe^{2+}), restoring the oxygen-carrying ability of blood.

2. Sepsis and Septic Shock:

• Methylene Blue has been investigated for its possible role in sepsis management. It may help regulate vasoplegia and improve hemodynamic in septic shock by inhibiting nitric oxide-mediated vasodilatation.

3. Vasoplegic Syndrome:

• In cardiac surgery, Methylene Blue is used to treat vasoplegic syndrome, a disease characterized by severe hypotension and vasodilatation following cardiopulmonary bypass.

4. Treatment with antibiotics:

• Methylene Blue can kill germs and parasites (antimicrobial activity). Researchers have looked into its possible use as an extra treatment for diseases like malaria and urinary tract infections.

5. Neuroprotection in diseases that damage nerve cells:

• Methylene Blue is being studied for its neuroprotective benefits because it is an antioxidant and can improve mitochondrial function. It could help lower reactive stress and protect neurons from damage in diseases like Parkinson's and Alzheimer's.

6. Disorders of the mind:

• Methyl blue has shown promise as an extra treatment for some mental illnesses, like bipolar disorder and sadness. It is thought to change neurotransmitter systems and make mood stability better.

7. In dermatology:

• In dermatology, Methylene Blue is applied to the skin to treat cyanosis caused by methemoglobinemia, wounds, and some skin diseases.

8. Instillations in the urinary tract and bladder:

• Methylene Blue can be injected into the bladder or urinary stream to treat conditions like interstitial cystitis or to help endoscopic surgeons find openings in the ureter.

9. Therapy with antioxidants:

• Methylene Blue may be useful as an antioxidant for lowering reactive stress caused by a number of diseases, which could improve the health and function of cells.

These medical uses for Methylene Blue show how useful it is in a wide range of situations, from treating serious injuries and illnesses right away to managing long-term conditions and exploring new treatments in many different areas of medicine.

Chapter Eight

Comparative analysis with alternative treatments

When thinking about how Methylene Blue can be used as a medicine, it's helpful to look at how it compares to other treatments in different medical areas. Here's a look at how Methylene Blue stacks up against some other widely used treatments:

1. Being methemoglobinic

Methylene Blue vs. Other Options:

- **Methylene Blue:**

- It works by giving electrons to turn methemoglobin into hemoglobin.

- The usual dose is 1-2 mg/kg given through an IV.

- It works very well, especially for acquired methemoglobinemia.

- **Alternative treatment:**

- High-Dose Vitamin C: Helps break down things, but not as well as Methylene Blue.

- Exchange Transfusion: This is used in the worst cases, especially with newborns, but it is more invasive and comes with more risks.

2. Getting Sepsis or Septic Shock

Methylene Blue vs. Other Options:

- **Methylene Blue:**

• Works by stopping nitric oxide synthase, which lowers dilatation.

• Use: It is usually used along with other treatments.

- **Alternative treatment:**

As a first line of defense against low blood pressure in septic shock, vasopressors like nor are epinephrine and epinephrine used.

- Corticosteroids: These are used to treat septic shock that won't go away, but long-term use endangered.

Combination treatment with hydrocortisone and fludrocortisones for septic shock that doesn't respond to vasopressors.

3. Neuroprotection in diseases that damage neurons

Methylene Blue vs. Other Options:

- **Methylene Blue:**

- Mechanism: It improves the performance of mitochondria and lowers oxidative stress.

- Status of the research: Looks good in animal tests and early-stage human studies.

- **Different types of treatment:**

- Ant cholinesterase inhibitors, like Donepezil and Rivastigmine, raise the amounts of neurotransmitters in people with Alzheimer's disease.

- Dopaminergic Agents, like Levodopa: This is the standard treatment for Parkinson's disease because it fixes a lack of dopamine.

- Antioxidants, like Vitamin E and Coenzyme Q10, are used to fight oxidative stress, but their effects vary.

4. Treatment with antibiotics

Methylene Blue vs. Other Options:

- **Methylene Blue:**

• Action: It messes with the DNA and cell walls of microbes.

• Use: Usually taken along with regular medicines.

- **Different types of treatment:**

Antibiotics, like Trimethoprim-Sulfamethoxazole and quinolones, are the first choice for treating bacterial diseases.

- Antiparasitics, like Artemisinin and Chloroquine, are the most common way to treat malaria.

5. Syndrome of Vasoplegic Methylene Blue vs. Other Options:

- **Methylene Blue:**

- It works by lowering the hypertension caused by nitric oxide.

- Dosage: 1-2 mg/kg is the usual amount given.

- **Different types of treatment:**

- Vasopressors, like phenylephrine and vasopressin,

are common first-line treatments for low blood pressure.

- Corticosteroids and immunoglobulins: These are sometimes used to change how the immune system reacts.

6. Mental Health Problems

Methylene Blue vs. Other Options:

- **Methylene Blue:**

- Mechanism: Changes in the potential of the dopamine and serotonin systems.

Investigational, showing some promise in treating sadness that doesn't respond to other treatments.

- **Different types of treatment:**

• Antidepressants (SSRIs, SNRIs) are the first choice for treating sadness.

• Mood stabilizers, like lithium and valproate, are used to treat bipolar illness.

Profiles of safety and effectiveness

1. Being methemoglobinic

- **What works:**

• **Very Effective:** Methylene Blue quickly lowers methemoglobin levels, which increases the body's ability to carry oxygen.

- **Quick Onset:** Most people feel better within minutes to hours of taking the medicine.

- **Safety:**

- **Safe for most people:** well-tolerated at appropriate doses.

- **Possible Side Effects:** People who don't have enough glucose-6-phosphate dehydrogenase (G6PD) may experience hemolysis if they take large amounts of the drug quickly.

- **Monitoring:** During treatment, it is best to keep a close eye on the amounts of oxygen saturation and methemoglobin.

2. Getting Sepsis or Septic Shock

- **What works:**

- **Adjunctive therapy:** This lowers dilatation to improve blood flow.

- **Not much evidence:** It works in different ways and is often used with other treatments, such as vasopressors.

- **Safety:**

- Be careful, because it could have bad effects, like serotonin syndrome if it's mixed with serotoninrgic drugs.

- **Hypotension Risk:** Can temporarily lower blood pressure, so it needs to be closely watched while being taken.

3. Neuroprotection in diseases that damage neurons

- **What works?**

- **Positive Preclinical Data:** It seems to have the ability to improve mitochondrial function and lower markers of neurodegeneration in animal models.

- **Clinical Trials:** Early-phase trials didn't give us a lot of information, so more study is needed to prove that it works in humans.

- **Safety:**

- **Generally Well-Tolerated:** Mild stomach problems and temporary changes in the color of the pee are common side effects.

- **Long-Term Use:** More research needs to be done on the safety of long-term use and in older people.

4. Treatment with antibiotics

- **What works?**

- **Add-on Role:** It makes normal antibiotics work better against microbes when used together.

- **Limited Use:** Only used in certain illnesses where antimicrobial resistance or other treatments are being thought about.

- **Safety:**

- **Generally Safe:** There is a low chance of bad effects when it is used correctly.

Depending on the dose, higher doses may cause stomach problems or very rare allergic responses.

5. Syndrome of Vasoplegic

- **What works?**

- **Sometimes useful:** helps keep blood pressure stable by stopping too much dilatation.

- **Varying Response:** How well it works depends on what's causing vasoplegia and how bad it is.

- **Safety:**

- **Be careful**, because this drug could cause temporary low blood pressure and other heart problems.

- **Monitoring:** Must be closely watched while being given, especially to patients who are very sick.

6. Mental Health Problems

- **What works?**

- **Research Use:** It has been studied as an extra treatment for people with depression and bipolar disorder who don't respond to other treatments.

- **Mechanism:** Changes in the potential of neurotransmitter systems, such as serotonin and dopamine.

- **Safety:**

Mixed Data: There isn't a lot of information on how safe and effective it is in the long run for psychiatric conditions.

As for side effects, they may include stomach problems, headaches, and very rarely, serotonin syndrome if used with other serotonin-related drugs.

Chapter Nine

Tests and trials in hospitals

Methylene Blue has been tested and studied in a lot of different medical areas, showing that it has a lot of different therapeutic uses. Some important clinical trials and studies that look at Methylene Blue in a variety of settings:

1. Being methemoglobinic

• A 2019 study in the Journal of Emergency Medicine looked at how well and safely Methylene Blue treats methemoglobinemia. The study emphasized how quickly and effectively it reverses cyanosis caused by methemoglobinemia.

• Clinical studies have shown that Methylene Blue is useful for both kids and adults, highlighting its role as a first-line treatment for methemoglobinemia caused by a number of things, such as medication-induced and inherited forms.

2. Getting Sepsis or Septic Shock

• Studies, some of which were published in Critical Care Medicine and Intensive Care Medicine, have looked into Methylene Blue's use as an extra treatment for septic shock. These studies looked at how it affected vasoplegia and blood flow, especially in people who didn't respond to regular vasopressor treatment.

• Clinical studies have looked at how safe and effective Methylene Blue is at lowering nitric oxide-mediated vasodilation and improving outcomes in septic patients. The results have been mixed, which means that more research is needed.

3. Neuroprotection in diseases that damage neurons

• Preclinical studies, like those in Neurobiology of Aging and Neuroscience Letters, have looked into how Methylene Blue protects neurons in models of Parkinson's and Alzheimer's illnesses. These studies point to possible ways that mitochondrial enhancement and antioxidant qualities might work.

• Pilot studies and small-scale trials in the early stages of clinical trials have looked at Methylene Blue's safety and preliminary effectiveness in slowing down cognitive decline and neurodegeneration in humans. This has set the stage for bigger trials.

4. Treatment with antibiotics

- Studies have looked at how well Methylene Blue kills different kinds of pathogens, such as bacteria that are resistant to multiple drugs and malaria parasites. In vitro studies and animal models have been used in these studies to show that it has the ability to be used with antibiotics to make them work better.

- Methylene Blue's effectiveness in treating certain infections and safety when mixed with common antibiotics have been tested in clinical studies. However, more clinical proof is needed before it can be used on a larger scale.

5. Syndrome of Vasoplegic

• Clinical studies and case reports show that Methylene Blue can help treat vasoplegic syndrome after heart surgery. This shows that it can balance blood flow by stopping excessive vasodilation.

• Trials have looked into the best way to give it and how well it works in critically sick patients. The results show that it can help lower the need for vasopressors and improve clinical outcomes, though it's important to keep in mind that it might have some side effects.

6. Mental Health Problems

- Randomized controlled trials (RCTs) and open-label trials have been used to look into Methylene Blue as an extra medication for people with depression and bipolar disorder who aren't responding to other treatments. The main focus of these research was on its possible role in changing neurotransmitter systems and making mood stabilization better.

- Methylene Blue's safety, tolerability, and long-term effectiveness in psychiatric conditions are still being studied. Trials are still going on to find out how it works and what therapeutic effects it has.

www.ingramcontent.com/pod-product-compliance
Lightning Source LLC
Chambersburg PA
CBHW071952210526
45479CB00003B/912